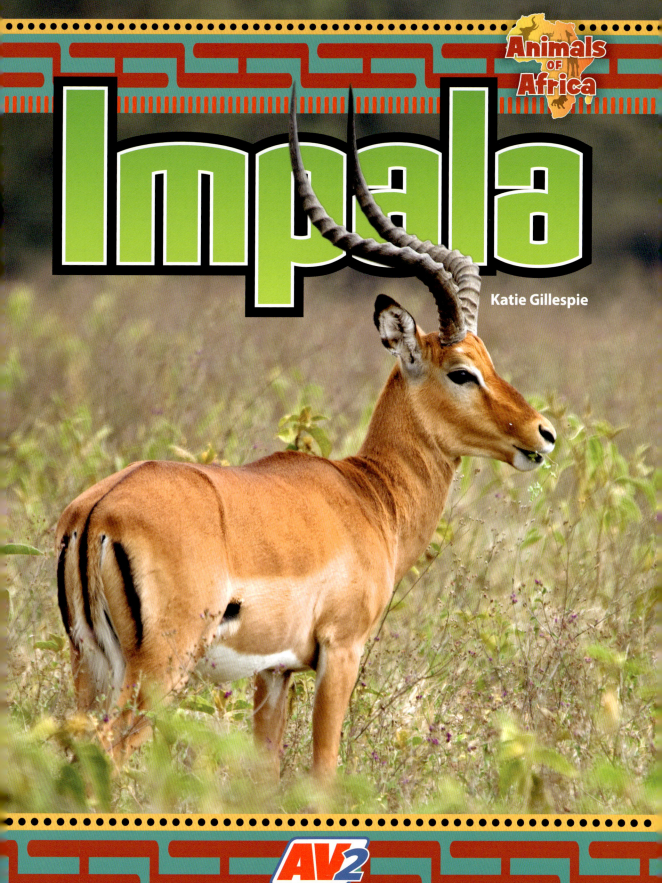

Impala

Katie Gillespie

Step 1
Go to **www.openlightbox.com**

Step 2
Enter this unique code

HRQEWCS2D

Step 3
Explore your interactive eBook!

CONTENTS

AV2 is optimized for use on any device

Your interactive eBook comes with...

Contents
Browse a live contents page to easily navigate through resources

Audio
Listen to sections of the book read aloud

Videos
Watch informative video clips

Weblinks
Gain additional information for research

Slideshows
View images and captions

Try This!
Complete activities and hands-on experiments

Key Words
Study vocabulary, and complete a matching word activity

Quizzes
Test your knowledge

Share
Share titles within your Learning Management System (LMS) or Library Circulation System

Citation
Create bibliographical references following the Chicago Manual of Style

This title is part of our AV2 digital subscription

1-Year Grades K–5 Subscription
ISBN 978-1-7911-3320-7

Access hundreds of AV2 titles with our digital subscription.
Sign up for a FREE trial at **www.openlightbox.com/trial**

Impala

CONTENTS

Meet the Impala

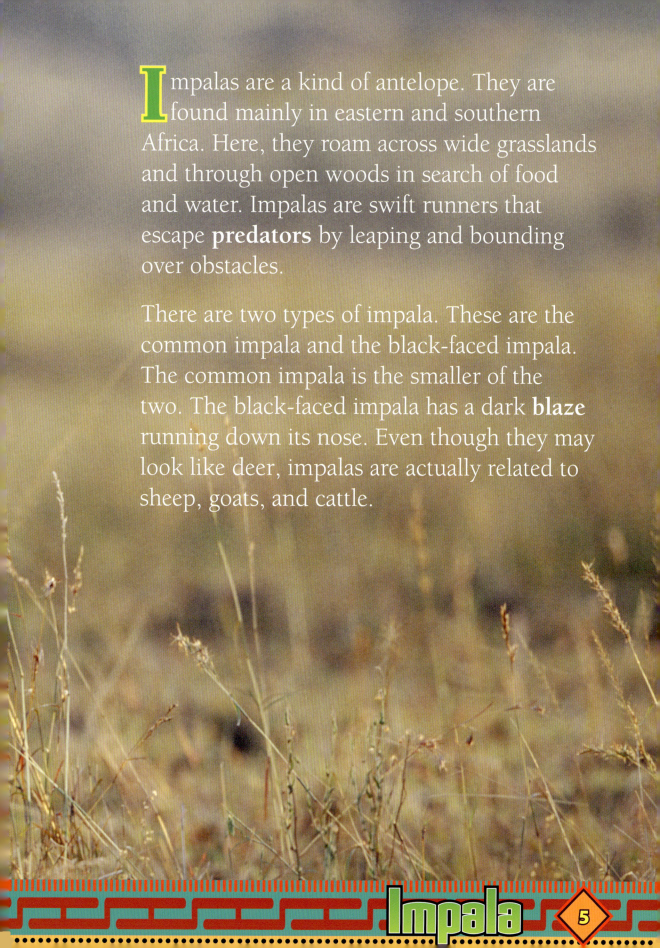

Impalas are a kind of antelope. They are found mainly in eastern and southern Africa. Here, they roam across wide grasslands and through open woods in search of food and water. Impalas are swift runners that escape **predators** by leaping and bounding over obstacles.

There are two types of impala. These are the common impala and the black-faced impala. The common impala is the smaller of the two. The black-faced impala has a dark **blaze** running down its nose. Even though they may look like deer, impalas are actually related to sheep, goats, and cattle.

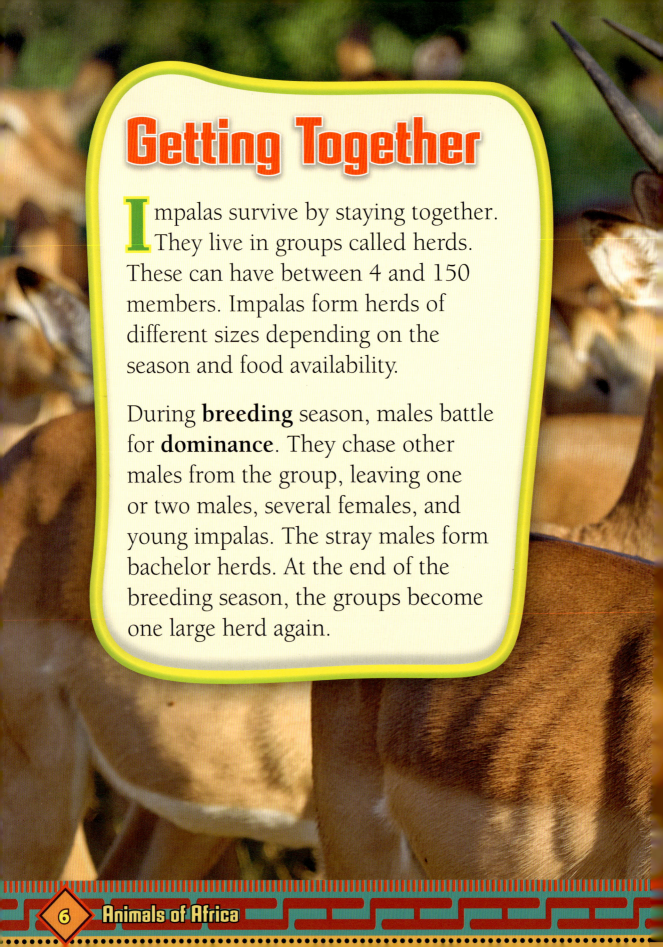

Getting Together

Impalas survive by staying together. They live in groups called herds. These can have between 4 and 150 members. Impalas form herds of different sizes depending on the season and food availability.

During **breeding** season, males battle for **dominance**. They chase other males from the group, leaving one or two males, several females, and young impalas. The stray males form bachelor herds. At the end of the breeding season, the groups become one large herd again.

Body Language

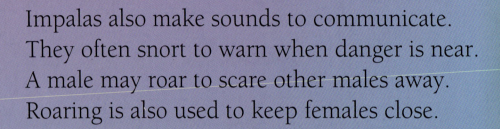

Impalas use their bodies to **communicate** and to defend themselves. When trying to attract a mate, they flick their tongues at one another. If a predator approaches, they may perform a kick with their back legs. This kick is used to confuse their pursuers.

Impalas also make sounds to communicate. They often snort to warn when danger is near. A male may roar to scare other males away. Roaring is also used to keep females close.

Grooming each other helps impalas establish and maintain relationships within the herd.

BIG FACT

The **roar** of a **male impala** can be **heard** more than **1 mile** (1.6 kilometers) **away**.

Growing Up

Baby impalas, or lambs, are typically born at the end of the **rainy season**. The weather is moderate at this time of year. This increases a lamb's chances of survival. Mother impalas can actually delay the birth of their lambs for up to a month to ensure that they are born after the rains.

When a mother impala has to be away from her lamb, she leaves it with other lambs in a crèche, or nursery group. Being in a group helps to keep the lambs safe. This is because predators are more likely to attack individuals.

Mother impalas give birth during the hottest part of the day, when predators are resting.

Comparing Life Spans

Elephant
70
years

Giraffe
25
years

Cheetah
12
years

Impala
15
years

A Closer Look

Impalas have **adapted** to their **habitats**. They have unique features that help them survive in their environment.

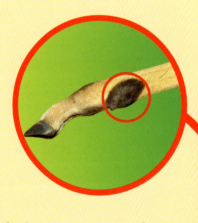

Coat

An impala's coat is dark on the back and light on the belly. This coloring helps **camouflage** the impala in grassy areas.

Back Legs

A special **gland** above an impala's back hooves leaves a scent on the ground. Other impalas follow this scent. This keeps the herd moving as a group.

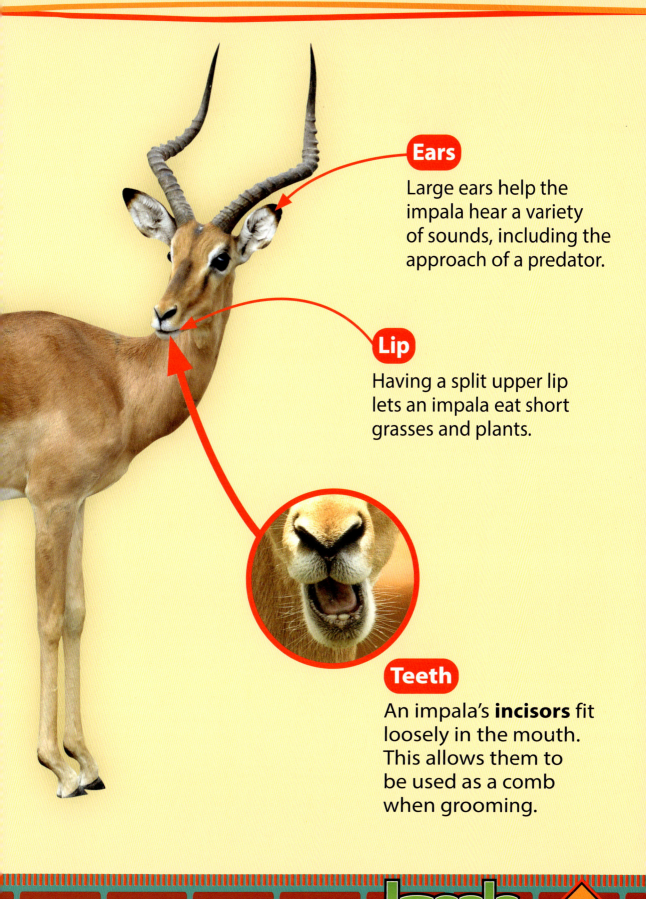

Ears

Large ears help the impala hear a variety of sounds, including the approach of a predator.

Lip

Having a split upper lip lets an impala eat short grasses and plants.

Teeth

An impala's **incisors** fit loosely in the mouth. This allows them to be used as a comb when grooming.

On the Curve

Male impalas are known for their curved horns. These horns take many years to reach their full length. They can grow to be 36 inches (91 centimeters) or longer, making them the largest antelope horns in east Africa. An impala's horns are strong, flexible, and hollow at the base.

Male impalas use their horns to defend themselves and to fight for dominance within a herd. The curved shape of the horns protects an impala's head when fighting. Female impalas do not have horns.

BIG FACT

An impala's **horns can grow** to be as long as its **body height**.

Leaps and Bounds

The impala's survival depends on its ability to escape predators. Impalas can reach speeds of up to 50 miles (80 km) per hour. They can also leap a distance of 33 feet (10 meters) and as high as 10 feet (3 m).

When fleeing, an impala will jump from one side to another, bumping up against other herd members. The herd will then scatter in a way that appears disorganized in order to confuse predators. While running, impalas keep together, soaring over bushes and other obstacles.

Cheetah
60 mph (97 kph)

Impala
50 mph (80 kph)

FINISH

Human
27 mph (43 kph)

Top Running Speeds

START

Daily Diet

Impalas are herbivores, which are animals that only eat plants. Impalas change their eating habits to fit the season. During the rainy season, they spend much of their time **grazing** on fresh grass. When the dry season arrives, impalas become browsers. This means that they eat leaves and shoots.

The ability to both graze and browse has made impalas highly adaptable. It helps them live in a range of habitats. Impalas can survive in areas where **livestock** has stripped the land of grass.

Impala herds feed together, with a few standing guard to watch for predators.

Impalas in Danger

Impalas drink water every day. They need constant access to water, making habitat loss in Africa a major concern. The building of roads and farms creates barriers for impala herd movement, and cuts off access to food and water.

There are more than 2 million impalas in nature. Roughly half the population lives on private land. Outside of protected areas, it can be difficult for herds to move freely in search of resources.

Impalas share their water sources with a number of other animals, including elephants.

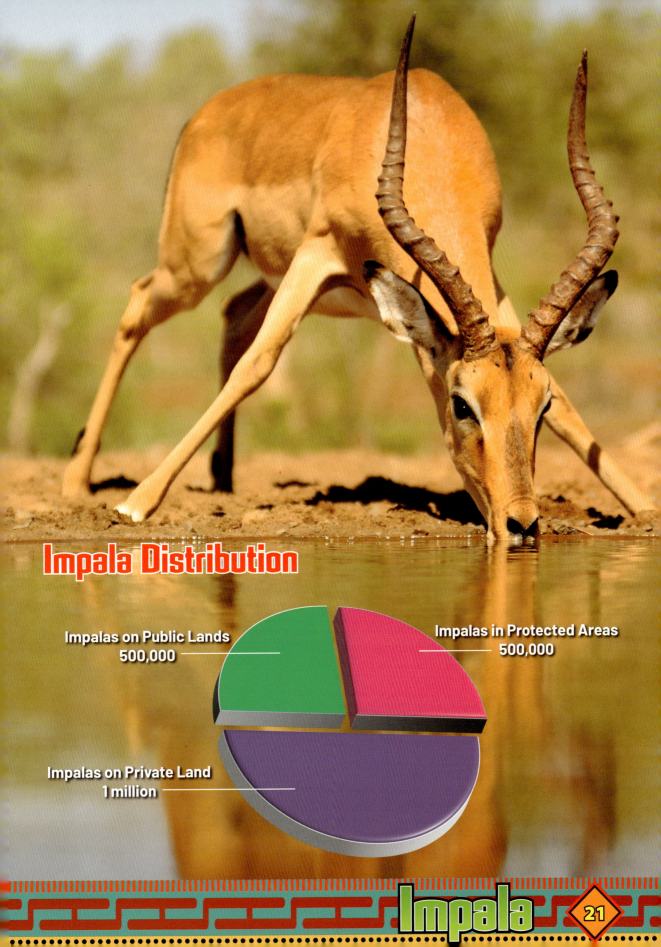

Impala Distribution

Impalas on Public Lands
500,000

Impalas in Protected Areas
500,000

Impalas on Private Land
1 million

Impala Quiz

1
What are the two types of impala?

2
How many impalas can be in a herd?

3
How far away can an impala's roar be heard?

4
What are baby impalas called?

5
How long can an impala's horns be?

6
How high can an impala leap?

7
What do herbivores eat?

8
How many impalas are left in nature?

ANSWERS 1. Common impala and black-faced impala **2.** Between 4 and 150 **3.** More than 1 mile (1.6 km) **4.** Lambs **5.** 36 inches (91 cm) **6.** As high as 10 feet (3 m) **7.** Plants **8.** More than 2 million

Key Words

adapted: changed to suit the environment

blaze: a colored stripe down the face of an animal

breeding: coming together to produce offspring

camouflage: to help an animal blend in with its environment

communicate: to exchange information, news, or ideas

dominance: having status or power over others

gland: a part of the body that makes one or more substances

grazing: feeding on growing grass

habitats: the places where animals live, grow, and raise their young

incisors: the sharp teeth at the front of the mouth

livestock: animals kept on a farm for use or profit

predators: animals that hunt other animals for food

rainy season: a time of year when most of an area's rain falls

Index

Get the best
of both worlds.

AV2 bridges the gap between print and digital.

The expandable resources toolbar enables quick access to content including **videos**, **audio**, **activities**, **weblinks**, **slideshows**, **quizzes**, and **key words**.

Animated videos make static images come alive.

Resource icons on each page help readers to further **explore key concepts**.

Published by Lightbox Learning Inc.
276 5th Avenue, Suite 704 #917
New York, NY 10001
Website: www.openlightbox.com

Library of Congress Cataloging-in-Publication Data

Names: Gillespie, Katie, author.
Title: Impala / Katie Gillespie.
Description: New York, NY : Lightbox Learning Inc., [2023] | Series: Animals of Africa | Includes index. | Audience: Grades 2-3
Identifiers: LCCN 2021055130 (print) | LCCN 2021055131 (ebook) | ISBN 9781791144227 (library binding) | ISBN 9781791144234 (paperback) | ISBN 9781791144241
Subjects: LCSH: Impala--Africa--Juvenile literature.
Classification: LCC QL737.U53 G539 2023 (print) | LCC QL737.U53 (ebook) | DDC 599.64/6096--dc23/eng/20211110
LC record available at https://lccn.loc.gov/2021055130
LC ebook record available at https://lccn.loc.gov/2021055131

Printed in Guangzhou, China
1 2 3 4 5 6 7 8 9 0 26 25 24 23 22

072022
101121

Project Coordinator: Heather Kissock
Designer: Terry Paulhus

Photo Credits
Every reasonable effort has been made to trace ownership and to obtain permission to reprint copyright material. The publisher would be pleased to have any errors or omissions brought to its attention so that they may be corrected in subsequent printings. Lightbox Learning acknowledges Getty Images, Alamy, Minden Pictures, and Shutterstock as its primary image suppliers for this title.